ULTIMATE SUPERCARS

LAMBORGHINI AVENTADOR

By Craig Ellenport

Kaleidoscope
Minneapolis, MN

The Quest for Discovery Never Ends

This edition first published in 2021 by Kaleidoscope Publishing, Inc.

No part of this publication may be reproduced in whole or in part without written permission of the publisher.

For information regarding permission, write to
Kaleidoscope Publishing, Inc.
6012 Blue Circle Drive
Minnetonka, MN 55343

Library of Congress Control Number
2020936096

ISBN
978-1-64519-264-0 (library bound)
978-1-64519-332-6 (ebook)

Text copyright © 2021 by Kaleidoscope Publishing, Inc. All-Star Sports, Bigfoot Books, and associated logos are trademarks and/or registered trademarks of Kaleidoscope Publishing, Inc.

Printed in the United States of America.

FIND ME IF YOU CAN!

Bigfoot lurks within one of the images in this book. It's up to you to find him!

TABLE OF CONTENTS

Chapter 1: The Record Holder 4

Chapter 2: From Tractors to Supercars 10

Chapter 3: Built for Speed 16

Chapter 4: Behind the Wheel 24

Beyond the Book 28

Research Ninja 29

Further Resources 30

Glossary 31

Index 32

Photo Credits 32

About the Author 32

Chapter 1
The Record Holder

The video was less than seven minutes long. Jenny saw it and was hooked. She couldn't believe it. The Lamborghini Aventador SVJ supercar had just set a new record.

The Nürburgring race track in Germany is famous in the car world. Drivers bring their best race cars to the track. They want to see how fast they can drive one 16.2-mile (26-km) lap.

On July 28, 2018, the Lamborghini Aventador SVJ finished a lap at Nürburgring in 6 minutes and 44 seconds. It was the fastest time in the history of the track!

SVJ in the Aventador's name stands for *Superveloce Jota*. That is Italian for "super fast race car." They sure chose the right name!

Jenny watched the Aventador SVJ speed around Nürburgring. She had one thought: "I need to drive this car!"

PARTS OF A LAMBORGHINI AVENTADOR

Air slots create downforce

LED headlights

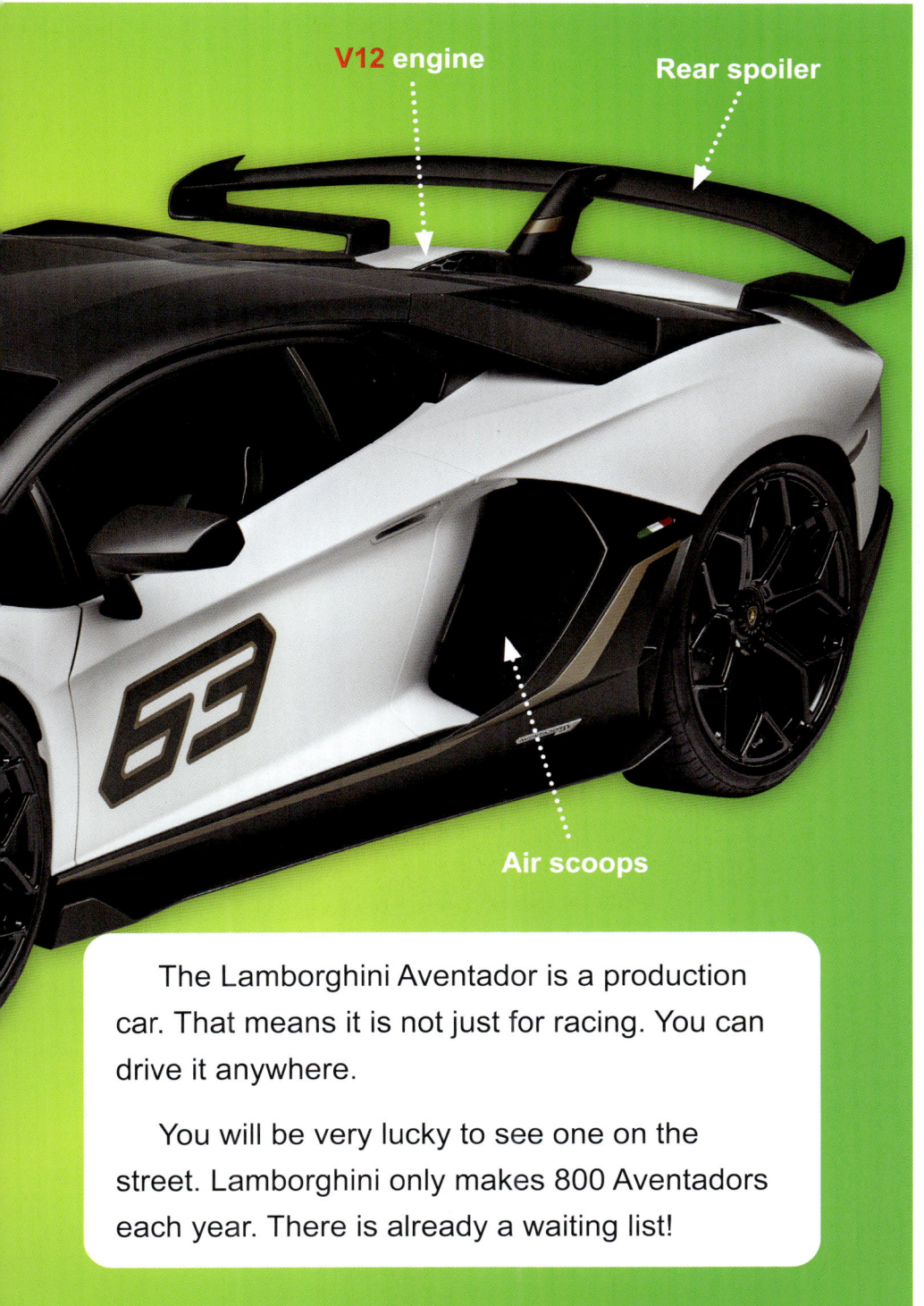

V12 engine

Rear spoiler

Air scoops

The Lamborghini Aventador is a production car. That means it is not just for racing. You can drive it anywhere.

You will be very lucky to see one on the street. Lamborghini only makes 800 Aventadors each year. There is already a waiting list!

Jenny has a plan. She lives near Las Vegas, Nevada. There is a place near Las Vegas called Exotics Racing. They have a race track and more than 50 sports cars. People can pay money to drive laps around the track. Exotics Racing has 11 Lamborghinis! Jenny prepares for her adventure. She reads about the Aventador on the Exotics Racing website.

FUN FACT
The main colors of the Aventadors come from the flag of Italy.

Romain Thieven is one of the owners of the business. He knows all about racing. He was a professional race car driver in France.

On the Aventador page, Thieven talks about the car's big engine. He says it is the most powerful car that Lamborghini makes. There is one thing Thieven says that Jenny really likes:

"The Aventador has only one goal in life, to go fast!"

Chapter 2
From Tractors to Supercars

Ferruccio Lamborghini grew up on a farm in Italy. He became good at fixing the farm equipment. He also liked to drive the tractor.

When he was a teenager, his father sent him to school to learn how to be a mechanic. After school, Ferruccio started his own tractor company.

Lamborghini's company made great tractors. He soon became a wealthy businessman. His company made big, slow tractors. But when he wasn't working, Lamborghini liked to drive small, fast sports cars.

At the time, the finest maker of sports cars in Italy was Ferrari. Ferruccio drove a Ferrari, but he thought he could make better cars.

This is a classic 1966 Lamborghini 400 GT.

FUN FACT
Lamborghini still sells tens of thousands of tractors a year!

Lamborghini wanted to build the best sports car in the world. He wanted his car to be better than a Ferrari. So he hired the best people to help him.

He hired Giotto Bizzarrini to build his engine. Bizzarrini used to build engines for Ferrari. Lamborghini hired two talented young engineers to help build the body of the car.

In 1964, Lamborghini's car company showed off its first super sports car, the 350 GT.

ANIMAL LOGOS

Lamborghini needed a logo for his car company. He wanted to compete with Ferrari. Ferrari's logo was a prancing horse. To show that he was part of the "stable" of great cars, Lamborghini chose a bull for his logo. Today, the Lamborghini factory has a nickname. It is called "The House of the Raging Bull."

This 1965 350 GT is taking part in a car rally in Italy.

Lamborghini continued to build super sports cars. Every new model had a stronger engine. Every new model was faster.

The company put out the Miura and the Espada. The Countach was another famous Lamborghini. The Huracan was named for a hurricane! All the cars were

AVENTADOR: THE BRAVEST BULL

Ferruccio Lamborghini visited a bull ranch before starting his car company. He was amazed by the beauty and the strength of these giant beasts.

Many of the Lamborghini car models are named for bulls. Aventador was a famous bull from Spain. It competed in the bullfights in Zaragoza, Spain. In 1993, Aventador was named the bravest bull of the year.

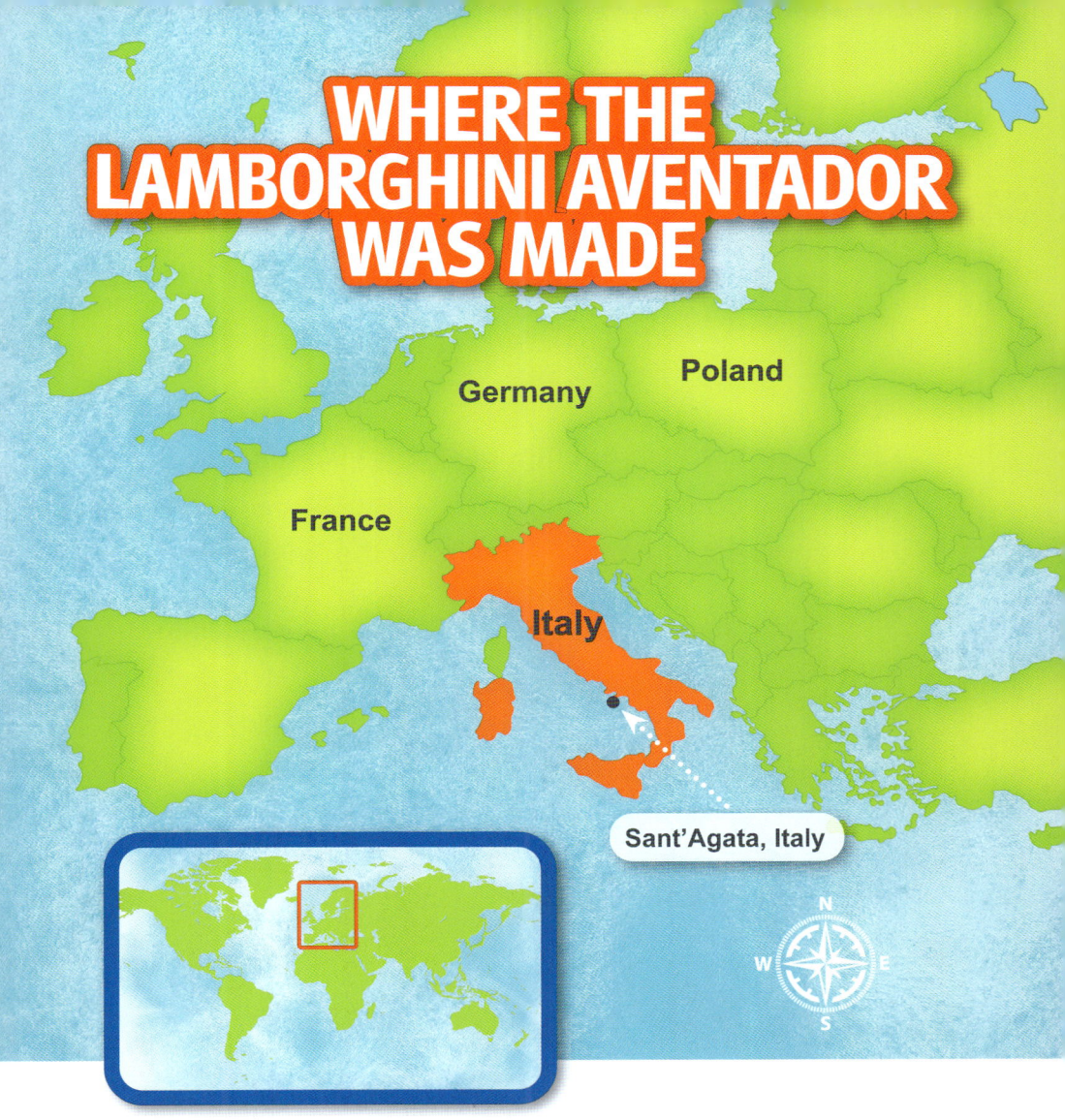

WHERE THE LAMBORGHINI AVENTADOR WAS MADE

beautiful and fast. And all of them were hard to get. The company only made a small number of each.

In 2011, Lamborghini put out a new style of car. This was the first Aventador. Lamborghini comes out with a newer version of the Aventador every year. The 2020 Aventador is stronger and faster than ever!

Chapter 3
Built for Speed

The Lamborghini Aventador has a V12 engine. The first model's engine had 700 **horsepower**. Horsepower measures how strong the engine is.

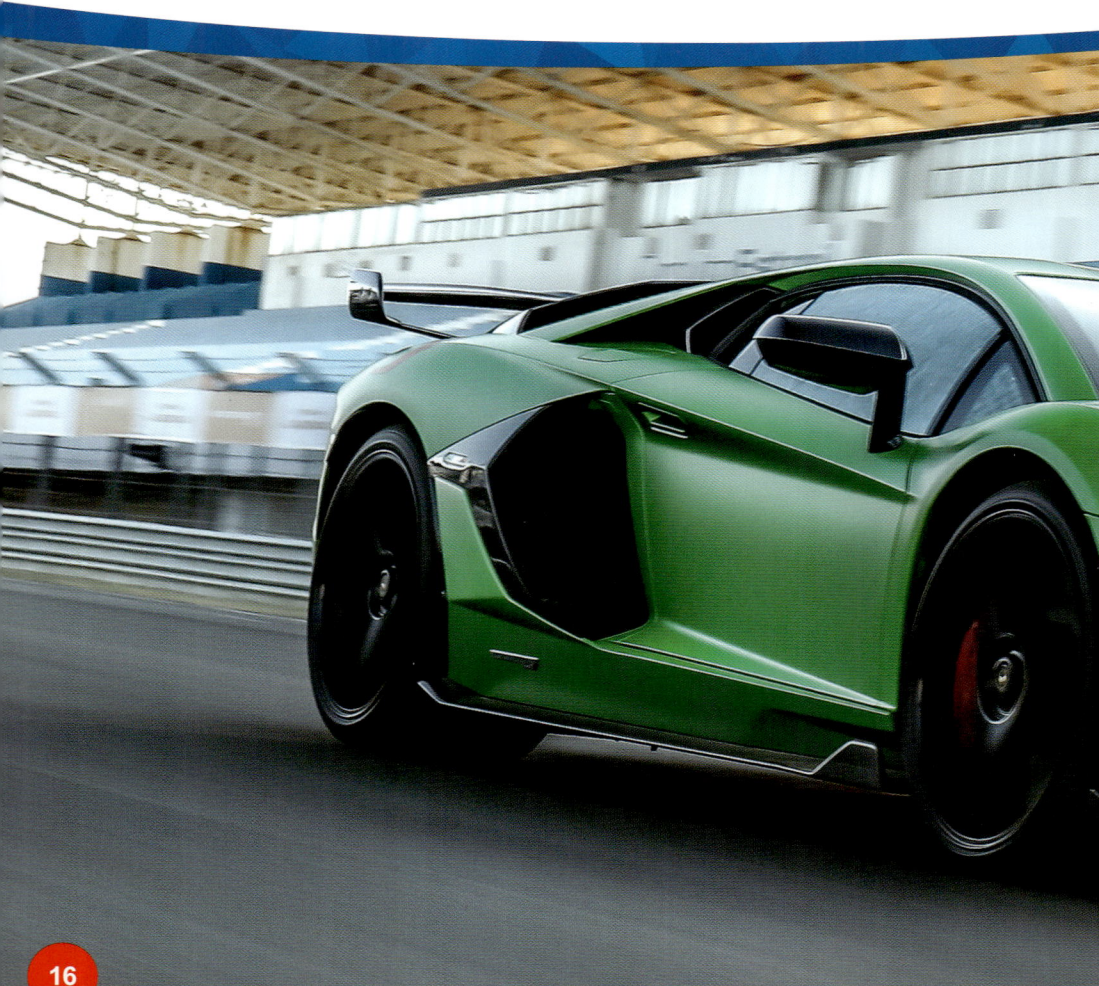

Most cars you see on the street have less than 200 horsepower.

The 2020 Aventador is even stronger. It has a 770 horsepower engine. This gives the car great **acceleration**. The Aventador can go over 200 miles (321 km) per hour.

The Aventador is low to the ground. It has good **aerodynamics**. That prevents wind from slowing it down. There is a bar on the back of the car. This is called a **wing**. When air hits the wing, it creates downforce. The air pushes down on the back of the car. This helps the tires grip the road. The car goes faster. It makes smoother turns.

The body of the car is made of something called **carbon fiber**. Carbon fiber is stronger than steel. It is also lighter than steel. This is another reason why the Aventador drives faster than other cars.

See how the shape of the Aventador cuts through the air.

SCISSOR DOORS

The Aventador looks different than most cars. When the doors are open, the Aventador *really* looks different.

The Aventador has "scissor doors." They don't open outward like most cars. Instead, the back of the door opens straight up. When both doors are open, they are sticking straight up. They look like scissors.

When the scissor doors open and close, the Aventador looks like a spaceship!

THE LAMBORGHINI AVENTADOR
IN DETAIL

Height: 3 feet, 8.7 inches (1.13 m)

Width: 6 feet, 8 inches (2.03 meters)

LENGTH: 15 feet, 6 inches (4.8 m)

WEIGHT: 3,386 pounds (1,536 kg)

TOP SPEED: 218 miles per hour (350 kph)

TIME FROM 0-60 MPH: 2.9 seconds

COST: $426,000 (United States)

The Aventador's engine is located in the back of the car. Having the engine in the back of the car is better for downforce. The weight of the engine pushes the car down toward the road. Air flows more smoothly over the car's body. That air flow also helps create downforce.

Going on a trip in your Lambo? There is storage space under the front hood. But don't take too much. There is not as much room as in most passenger cars!

There are two types of Aventadors. The **coupe** has a hard top. The **roadster** is a convertible. But it is not like most other convertibles. Most convertibles have a soft top that opens. The Aventador has a hard

top. But it has two panels that can be removed. There is enough space under the hood to store the panels. The driver can put the panels back on if it starts to rain!

This shows the two panels that make up the roof of the convertible. Pop them off and store them to feel the wind in your hair as you drive!

Chapter 4
Behind the Wheel

The day finally arrived for Jenny's big adventure. She got to Exotics Racing very early. She spent the extra time watching other drivers take their turn on the track.

Jenny signed in at the front desk. She was given lessons on how to drive on the race track. Then she was given a helmet. She paid to drive five laps around the track. That was a total of six miles. There is no speed limit on the track. Jenny can drive as fast as she wants!

Helmet on, Jenny opened the scissor door and slid into the driver's seat. The steering wheel was wrapped in leather. The dashboard had two video screens. Jenny felt like she was at the controls of a jet!

Jenny was ready to drive. She started the car and the engine roared. It was like nothing she had ever heard. The Aventador is known for having a loud engine. A loud engine is good. It tells the driver the engine is strong.

Jenny put the car in drive. She gently pressed her right foot down on the gas pedal. The Aventador took off!

Jenny was going 120 miles per hour. She couldn't believe it! Even though the car was going so fast, it was hard to tell. It was the smoothest car ride she ever took. She was gliding through the air.

LAST OF ITS KIND

The 2020 Aventador is the last of its kind. Lamborghini's next Aventador will be a hybrid car. A hybrid has two different engines. One runs on gas. The other is electric. These cars save energy. They are better for the environment.

A hybrid Aventador will not be as fast. But it will still look like a supercar!

Her ride was over in just a few minutes. Jenny stopped the car and got out. Her heart was still racing. Someone asked her how she liked the Aventador. Jenny smiled.

"It's like riding a wild bull," she said.

FUN FACT
A model maker made a 1/8-size Aventador model from gold and other metals. It sold for $4.8 million!

BEYOND
THE BOOK

After reading the book, it's time to think about what you learned. Try the following exercises to jumpstart your ideas.

RESEARCH

FIND OUT MORE. Where would you go to find out more about your favorite cars? Find out what company makes the car and locate its website. What information do the companies provide? What other sources of car information can you find?

CREATE

GET ARTISTIC. Cars start with creative artists and designers. Time for you to take a shot! Get art materials and create a great, new car. Will you make it a sports car? A sedan? A race car? What colors will you paint it? What features can you give it? Let your imagination go for a spin!

DISCOVER

DIG DEEPER. Lamborghini started out making tractors. What if he had decided to stick to tractors . . . but still wanted to race. Design a super-fast, high-tech racing tractor. What engine would it have? How would the driver be protected? What sort of extra features can you add?

GROW

GO TO A CAR SHOW. Car shows are a great way to see lots of cool cars up-close. Check your local events calendar, or ask at a car dealer for upcoming events. You can find shows of old cars and new cars, sports cars and classic cars. Go to a show and find a new favorite car to love!

RESEARCH NINJA

Visit **www.ninjaresearcher.com/2640** to learn how to take your research skills and book report writing to the next level!

RESEARCH

DIGITAL LITERACY TOOLS

SEARCH LIKE A PRO
Learn about how to use search engines to find useful websites.

FACT OR FAKE?
Discover how you can tell a trusted website from an untrustworthy resource.

TEXT DETECTIVE
Explore how to zero in on the information you need most.

SHOW YOUR WORK
Research responsibly— learn how to cite sources.

WRITE

GET TO THE POINT
Learn how to express your main ideas.

PLAN OF ATTACK
Learn prewriting exercises and create an outline.

DOWNLOADABLE REPORT FORMS

Further Resources

BOOKS

Cruz, Calvin. *Lamborghini Aventador.* Mankato, MN: Bellwether Media, 2016.

Garstecki, Julia. *Lamborghini Aventador.* Mankato, MN: Black Rabbit Books, 2018.

McKenna, A.T. *Lamborghini.* Minneapolis, MN: Abdo Publishing, 2000.

WEBSITES

Factsurfer.com gives you a safe, fun way to find more information.

1. Go to www.factsurfer.com.
2. Enter "Lamborghini Aventador" into the search box and click 🔍
3. Select your book cover to see a list of related websites.

Glossary

acceleration: an increase in speed.

aerodynamic: having a shape that can move through air quickly.

carbon fiber: a strong, lightweight material made from woven pieces of carbon.

coupe: a car with a hard roof and two doors.

horsepower: a unit of power that equals the work done in lifting 550 pounds one foot in one second.

hybrid: a mix of two different things, such as electric and gasoline motors in one car.

roadster: a car with an open top and two seats.

V12: an engine with 12 cylinders mounted on the crankshaft.

wing: a part on the back of a car that helps the car grip the road.

Index

350 GT, 12, 13
Bizzarrini, Giotto, 12
bulls, 13, 14, 27
convertible, 22, 23
Countach, 14
engine, 7, 9, 12, 14, 16, 17, 21, 26
Exotics Racing, 8, 24
France, 9, 15
Germany, 4, 15
Huracan, 14
hybrid car, 26

Lamborghini, Ferruccio, 10, 12, 13, 14, 15
Las Vegas, 8
logo, 13
Miura, 14
Nurburgring, 4
scissor doors, 19, 24
Spain, 14
Superveloce Jota, 4
Thieven, Romain, 9
tractors, 10, 11

PHOTO CREDITS

The images in this book are reproduced through the courtesy of: Courtesy Lamborghini: 4, 6, 8, 12, 16, 18, 19, 20, 21, 22, 25, 27. Shutterstock: FernandoV 10; Fernando Cortes 14; MakeSushi1 17. Wikimedia: Thomas Vogt 11.
Cover: RomanSt-Photographer/Shutterstock (car); Aleksandra H. Kossowska/Shutterstock (background, top); zhao jiankang/Shutterstock (background, bottom).

About the Author

Craig Ellenport is an award-wining journalist, author and editor from Massapequa, New York. He has authored numerous books and articles on subjects ranging from football and tennis to social media and health.